FIELD GUIDES
FOR KIDS

BIRDS

Tracy Abell

Abdo Reference

An Imprint of Abdo Publishing | abdobooks.com

CONTENTS

Birds are animals with backbones, and they are the only vertebrates with feathers and wings. Feathers and wings are very important to these animals for different reasons:

- Birds need their feathers to survive. Feathers help birds fly and swim, protect their skin from injury, and also trap heat to keep them warm. Brightly colored feathers help birds get mates. Feathers can also help a bird blend into its surroundings so predators can't see it.

- Different birds have different wing sizes and shapes. Many large birds have long wings that let them glide without flapping. Smaller birds often have short wings that let them fly in and out of vegetation. Even though all birds have wings, some cannot fly. However, their wings are still important. For example, ostriches use their wings for balance when they run.

OTHER IMPORTANT BIRD PARTS

- Most birds have hollow bones. Humans have cartilage between their bones, but birds' bones are fused together, which makes the bones strong. That strength, along with the air inside the bones, makes it easier for birds to fly.

- Most birds have nine air sacs located in their bodies and hollow bones. The sacs store air and help push air through the lungs.

- A bird's nose is on its beak, otherwise known as its bill. Birds use their beaks to find food, tear food into smaller pieces, and drink water. Birds also use them to attack predators and to communicate.

IDENTIFYING BIRDS

There are an estimated 18,000 bird species in the world. Birds from the same species might look different depending on what region they live in. With so many birds out there, it can be hard to identify them. It gets even more complicated when males and females have different color patterns. Males are usually brighter than females. Young birds also look different from older birds. And birds' feathers often look different during breeding season than during nonbreeding months.

HOW TO USE THIS BOOK

Tab shows the bird category.

The bird's common name appears here.

BIRDS OF PREY

RED-TAILED HAWK
(BUTEO JAMAICENSIS)

This paragraph gives information about the bird.

Red-tailed hawks are the most common hawk species in ... erica. They soar above open fields, turning in slow ... om below, their red tails are visible. The underside ... ings is light except for the dark wing tips. They ... ch on telephone poles and wires, scanning the ... r prey. Pairs who have mated usually stay together ... dies.

FUN FACT

Red-tailed hawks court by making wide circles as they fly ...gh in the sky. Then, ...ey sometimes grab ...ch other, clasping ...alons, and spiral ...oward the ground ...efore letting go.

Fun Facts give interesting information related to birds.

HOW TO SPOT

Color Pattern: Their upperparts are dark brown and underparts are pale. The tail has a dark rust coloring above and is lighter below.

Size: Around 17.7 to 22.1 inches (45 to 56 cm) long

Range: North America

Habitat: Open areas including deserts, grasslands, fields, and roadsides

Diet: Small mammals, birds, and snakes

Images show the bird.

TURKEY VULTURE *(CATHARTES AURA)*

Turkey vultures have red heads that lack feathers, and they have pale bills. These birds fly with their wings raised in a V, wobbling from side to side. They usually fly at low altitudes while searching for carrion. When threatened, they hiss. They also defend themselves by vomiting stomach

HOW TO SPOT

Color Pattern: These birds have brown upperparts with darker brown on the neck and below.

Size: About 25.2 to 31.9 inches (64 to 81 cm) long; 4.4 pounds (2 kg)

Range: North America, Central America, and South America

Habitat: Open areas including farmland, forests, landfills, and roadsides

Diet: Dead mammals, reptiles, birds, and fish

TURKEY VULTURES DO ALL THE WORK

Unlike other birds of prey, turkey vultures have a powerful sense of smell. Other birds watch as turkey vultures find the food, then chase the vultures off when they see them circling a carcass. Eagles and coyotes get to eat first. Then condors eat. Last in line are ravens and the turkey vultures.

45

OSTRICH *(STRUTHIO CAMELUS)*

This flightless species is the largest living bird. An ostrich's small head rests on top of a long neck. Ostriches have powerful legs and feet with just two toes. These help them fight predators and run fast. When running to escape lions, they can reach speeds of 43 miles per hour (70 kmh).

HOW TO SPOT

Color Pattern: Males are mostly black and white, while females have more brown coloring. Both sexes have heads and necks with pinkish skin covered with downy feathers.

Size: 7 to 9 feet (2.1 to 2.7 m) tall; 220 to 350 pounds (100 to 160 kg)

Range: Africa

Habitat: Plains and semidesert savannas

Diet: Plants, fruit from shrubs, locusts, and grasshoppers

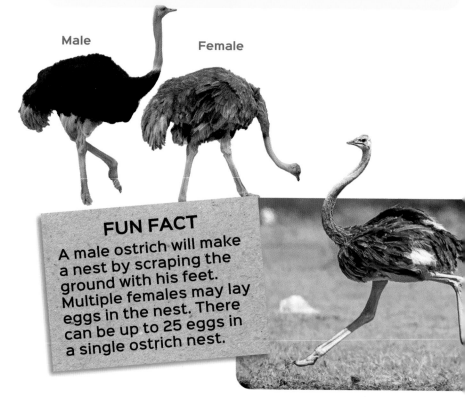

Male

Female

FUN FACT

A male ostrich will make a nest by scraping the ground with his feet. Multiple females may lay eggs in the nest. There can be up to 25 eggs in a single ostrich nest.

SOUTHERN CASSOWARY
(CASUARIUS CASUARIUS)

Southern cassowaries are large, shy birds. They scour the rain forest for fruit and are good swimmers. If threatened, these birds can get aggressive. They make a deep, booming sound which carries through the forest. Scientists believe the helmet on top of their heads helps amplify their booming voices.

HOW TO SPOT

Color Pattern: These birds have shaggy, black feathers, red wattles, dark blue and purple necks, and bare, blue skin on their heads.

Size: Up to 5.5 feet (1.7 m) tall; males weigh around 121 pounds (55 kg) and females weigh around 167 pounds (76 kg)

Range: Australia, New Guinea, and Indonesia

Habitat: Dense rain forests

Diet: Fallen fruit, snails, fungi, and some small mammals and reptiles

ANATOMY OF A SOUTHERN CASSOWARY

Southern cassowaries have a 5-inch-(12.7 cm) long claw on each foot. The helmet on their heads are hard on the outside. When provoked, they kick and sometimes jump on their victims. These birds may look fierce, but they are peaceful when left alone.

MALLARD *(ANAS PLATYRHYNCHOS)*

Mallards are the most common duck in North America. They have orange feet, round heads, and flat, wide beaks. Mallards are dabbling ducks, which means their heads go underwater to find food while their back ends remain above water.

Female

Male

FUN FACT

Female mallards quack, while male mallards make a quieter sound that is harsh and scratchy.

HOW TO SPOT

Color Pattern: Males have a green head, brown chest, black back end, and yellow bill. Females are spotted brown with a dark bill speckled with orange. Both males and females have a blue patch on their wings.

Size: 20 to 26 inches (50 to 66 cm) long

Range: North America

Habitat: Marshes, swamps, rivers, lakes, and city parks

Diet: Seeds, plants, earthworms, and grains

DABBLERS VS. DIVERS

Ducks are mostly divided into two categories: dabblers and divers. Dabbling ducks paddle on top of the water and skim the surface for food. They often dip their heads underwater. Divers swim low in the water. They dive for food, staying under for 10 to 20 seconds or longer.

PINK-EARED DUCK
(MALACORHYNCHUS MEMBRANACEUS)

Pink-eared ducks are also known as zebra ducks because of the white and black stripes on their sides. They have pink feathers near their eyes. Their large bills have grooves that help strain tiny plants and animals from the water. These small ducks feed together, circling head-to-tail in pairs or groups, rotating through the water. They are sociable and interact with other kinds of ducks. Pink-eared ducks often perch on logs and branches.

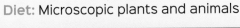

HOW TO SPOT

Color Pattern: These ducks have white sections on their faces, a dark patch over their eyes, and a gray bill.

Size: 14 to 18 inches (35 to 46 cm) long

Range: Australia

Habitat: Inland swamps, freshwater wetlands, and sewage ponds

Diet: Microscopic plants and animals

CANADA GOOSE *(BRANTA CANADENSIS)*

The Canada goose is the most widespread goose in North America. These large geese are seen in parks and on golf courses and lawns. They communicate with a deep honking sound and fly in *V* formations. They have long necks covered in black feathers, and their feet are black as well.

HOW TO SPOT

Color Pattern: Canada geese have a black head, white cheeks and chinstrap, and a brown back.

Size: 30 to 43 inches (76 to 109 cm) long; 6 to 20 pounds (2.7 to 9 kg)

Range: North America

Habitat: Waters, grassy fields, and grain fields

Diet: Grasses, berries, and seeds

FUN FACT

Canada geese are considered nuisances in some cities and suburbs because there are so many of them.

12

MAGPIE GOOSE
(ANSERANAS SEMIPALMATA)

Magpie geese are sometimes found in flocks of thousands. If a predator comes near, the entire flock of geese raises their heads. They will often fly straight up as they honk loudly. Magpie geese have long, black necks, orange legs and feet, and red skin on their faces. They also have a visible bump on their heads.

HOW TO SPOT

Color Pattern: Magpies have a white torso. Their backs, necks, breasts, and wings are black.

Size: 28 to 35 inches (71 to 89 cm) long; around 4.5 to 6.5 pounds (2 to 3 kg)

Range: Australia and New Guinea

Habitat: Shallow wetlands, swamps, and flooded grasslands

Diet: Grasses, seeds, and bulbs from reeds and rushes

BLACK SWAN *(CYGNUS ATRATUS)*

When the black swan isn't flying, it looks completely black. However, when it takes off into the air, viewers can see white feathers on its wing tips. Black swans have an orange-red beak. These swans shed their feathers after breeding and become flightless. During this period, they gather on the water in flocks of thousands.

HOW TO SPOT

Color Pattern: Black swans are mostly black with white wing feathers. Their eyes are red.

Size: Around 3.9 to 4.6 feet (1.2 to 1.4 m) long; 8 to 19 pounds (3.5 to 8.6 kg)

Range: Australia and Tasmania

Habitat: Large, open waters, flooded fields, and tidal mudflats

Diet: Aquatic plants and some insects

TUNDRA SWAN *(CYGNUS COLUMBIANUS)*

Tundra swans have long, white necks. Their beaks are black, usually with a yellow spot near the base. Their legs and feet are also black. When they fly, tundra swans' wings make a whistling sound, so they are also known as whistling swans. They often dip their heads underwater when looking for food. Once a male and female form a bond, they stay together year-round.

HOW TO SPOT

Color Pattern: All white

Size: 3.9 to 4.8 feet (1.2 to 1.5 m) long; about 8.4 to 23 pounds (3.8 to 10.4 kg)

Range: Canada and northern United States

Habitat: Lakes and ponds

Diet: Mostly plants, plus mollusks and crustaceans

FUN FACT
A male swan is called a cob and the female is a pen. Baby swans are called cygnets.

15

COMMON LOON *(GAVIA IMMER)*

Common loons are large diving birds with rounded heads and sharp bills. Loons only go ashore to mate and nest. They dive for fish and often swallow them while still underwater. In order to fly, loons need at least 30 yards (27 m) of open water. They run on the surface, flapping their wings to take off. Their colorings change at different times of the year.

Young loons have brown coloring.

HOW TO SPOT

Color Pattern: In the summer months, a common loon has a black head and bill, a black-and-white spotted back, and a white breast. During the rest of the year, it has a plain gray head, back, and bill. Its throat is white.

Size: 2 to 3 feet (0.6 to 0.9 m) long; 6.5 to 12 pounds (3 to 5.4 kg)

Range: Canada and United States

Habitat: Lakes and ponds

Diet: Fish, crustaceans, and snails

PIED-BILLED GREBE
(PODILYMBUS PODICEPS)

Pied-billed grebes are small, compact swimming birds with no tail. Sometimes, they float with just the upper half of their heads above water. They catch fish by diving or slowly submerging. They use their chunky bills to kill large crustaceans. Pied-billed grebes are rarely seen flying and spend a lot of time hiding in vegetation.

HOW TO SPOT

Color Pattern: Brown, with darker coloring on top and lighter beneath. During the breeding season, the bill is whitish with a black band. Otherwise, it is yellow brown.

Size: 12 to 15 inches (30 to 38 cm) long

Range: North America and Central America

Habitat: Small ponds and marshes

Diet: Crustaceans, small fish, and insects

FUN FACT
Pied-billed grebes eat their own feathers and also feed them to newly hatched chicks. Those feathers protect their intestines against sharp bits of the prey they eat.

ANHINGA *(ANHINGA ANHINGA)*

The anhinga is also known as the snakebird because it often swims with only its snake-like neck and head above water. It can swim underwater too, stabbing fish with its pointy bill. These birds are slim and have long tails.

HOW TO SPOT

Color Pattern: Adult males have black feathers with silvery-white streaks on their wings and backs. The necks and chests of females and juveniles are pale tan.

Size: 30 to 37 inches (75 to 95 cm) long; about 3 pounds (1.3 kg)

Range: Southern United States and Mexico

Habitat: Freshwater lakes and ponds; bays and lagoons along coasts

Diet: Small to medium-sized fish

Male

Female

FLIGHTLESS CORMORANT
(PHALACROCORAX HARRISI)

Like their name suggests, flightless cormorants cannot fly. They have short wings that help them balance as they jump from rock to rock. While swimming, they tuck in their wings and kick with strong legs. They dive deep to spear eels and octopuses with their sharp bills. A flightless cormorant's eyes are bright blue, its bill is brownish gray, and its legs and feet are black.

HOW TO SPOT

Color Pattern: Black on top and dark brown underneath
Size: 35 to 39 inches (90 to 100 cm) long; up to 11 pounds (5 kg)
Range: Galapagos Islands (Isabela and Fernandina)
Habitat: Lava shorelines and beaches
Diet: Eels and octopuses

FUN FACT
During their courtship dances, a pair of flightless cormorants entwine their necks. Then, they spin in a tight circle.

19

CHINSTRAP PENGUIN
(PYGOSCELIS ANTARCTICUS)

Chinstrap penguins have a black band that runs under their chins. Their bills and eyes are also black, and they have pale pink feet. Chinstrap penguins hunt for food in the water and go on land to breed, nest, and escape from predators such as leopard seals.

FUN FACT
Chinstrap penguins gather in groups known as colonies. One such colony includes 2.4 million penguins!

HOW TO SPOT

Color Pattern: A chinstrap penguin's head, back, and tops of flippers are black. Its face and underparts are white.

Size: 28 inches (71 cm) long; 6.6 to 11 pounds (3 to 5 kg)

Range: Islands in the Antarctic Ocean

Habitat: Rough, rocky locations

Diet: Krill and fish

WHY NOT FLY INSTEAD?

Penguins used to fly, but scientists believe they evolved to become better at swimming. Their flippers allow them to dive deep for food. For instance, an emperor penguin can hold its breath for more than 20 minutes as it dives 1,500 feet (450 m).

HORNED PUFFIN
(FRATERCULA CORNICULATA)

The horned puffin has a large, bright-yellow bill with an orange tip. Its eyes are dark and circled with red, and there's a black "horn" above each eye. Its feet are red orange. Horned puffins spend most of their lives on the open water. During breeding season they come on land to nest in cracks and burrows in sea cliffs.

HOW TO SPOT

Color Pattern: Upperparts are black, and its face and underparts are white.

Size: 15 inches (38 cm) long; 1.4 pounds (0.6 kg)

Range: North Pacific Ocean

Habitat: Ocean waters, sea cliffs, rocky islands, and grassy bluffs

Diet: Small fish, shrimp, and krill

21

BROWN PELICAN
(PELECANUS OCCIDENTALIS)

Brown pelicans are large birds with slender necks. Their bills are tan, brown, and red orange. When they dive headfirst for fish, brown pelicans tuck and turn their heads to protect their throats from the impact. Their throat pouches expand to trap fish, filling with more than 2 gallons (7.5 L) of water.

HOW TO SPOT

Color Pattern: These pelicans are gray brown with a yellow head and white neck.

Size: 39 to 54 inches (99 to 137 cm) long; around 4.5 to 11 pounds (2 to 5 kg)

Range: Southern coasts of the United States

Habitat: Breed and nest on barrier islands; rest on sandbars, breakwaters, and offshore rocks

Diet: Small fish

NORTHERN GANNET
(MORUS BASSANUS)

The northern gannet spends most of its life at sea. These birds form flocks of thousands that often dive for fish at the same time. Their calls and good vision help them not crash into each other. Northern gannets' dives are usually shallow, but they can go as deep as 72 feet (22 m). They use their wings and feet to swim for fish.

HOW TO SPOT

Color Pattern: White with black wing tips and a yellow head

Size: 37 to 43 inches (94 to 110 cm) long; around 5.6 to 8 pounds (2.5 to 3.6 kg)

Range: North America and Europe

Habitat: These birds live near open oceans and large bays. They nest on cliffs.

Diet: Fish, shrimp, and squid

FUN FACT
Northern gannets' eyes are adapted for diving. As soon as they hit the water, they can see as well as they do in flight.

MAGNIFICENT FRIGATE BIRD
(FREGATA MAGNIFICENS)

Magnificent frigate birds are tropical seabirds with very long wings. Their bills are long and hooked. They use their deeply forked tails to steer as they glide through the air, barely flapping their wings.

FUN FACT

Magnificent frigate birds will chase other birds until those birds throw up recently eaten food. They swoop to catch the food before it hits the water.

Male

HOW TO SPOT

Color Pattern: These birds are mostly black and have a grayish bill. Males have a red throat pouch that looks like a balloon when inflated. Females have a white chest.

Size: 35 to 45 inches (89 to 114 cm) long

Range: Coastlines of southern United States, Mexico, and the Caribbean

Habitat: Coasts, islands, coral reefs, low trees, and shrubs

Diet: Fish, plankton, crabs, and jellyfish

BLUE-FOOTED BOOBY
(SULA NEBOUXII)

The blue-footed booby is a large seabird with a long, pointed bill and wings. But it's the adults' blue feet that attract attention. Their mating dances involve the male lifting his feet and then bowing to the female. The brighter the blue, the better the male's chance at finding a mate. These birds will dive for fish from 80 feet (24 m) in the air, and can go as deep as 65 feet (20 m) underwater.

HOW TO SPOT

Color Pattern: The bird's upperparts are brown and underparts are white. The head is whitish, and the bill is dark gray.

Size: 32 to 34 inches (81 to 86 cm) long

Range: Western coasts of Central America and South America

Habitat: Rest and roost near shores and on rocks onshore

Diet: Males eat small fish; females eat larger fish and squid

BRIDLED TERN
(ONYCHOPRION ANAETHETUS)

Bridled terns have a forked tail, long wings, and a pointed black-gray bill. The top of their heads are black, and their foreheads are white. Bridled terns fly low over water. Instead of diving for food, they usually grab fish at the water's surface. They are mostly silent birds but sometimes make a soft *wheep* sound.

HOW TO SPOT

Color Pattern: Their backs are brown, and their undersides are white.

Size: 15 inches (38 cm) long

Range: Tropical seas

Habitat: Warm offshore ocean waters; nests on islands with rocks, caves, or bushes

Diet: Small fish, crustaceans, and insects

ARCTIC TERN MIGRATION

Every year, another species of tern migrates from one pole to the other. Arctic terns fly from Greenland in the north to the Weddell Sea in the south. The trip is around 56,000 miles (90,000 km)—the farthest known migration of any animal.

LAUGHING GULL
(LEUCOPHAEUS ATRICILLA)

Laughing gulls have black or reddish-black legs and long wings. The gull's call is a loud series of laughing notes. When a gull is defending a nest, its call can last several minutes. These birds visit landfills and parking lots looking for food. They sometimes eat the eggs of other birds.

HOW TO SPOT

Color Pattern: Upperparts are gray and underparts are white. In the summer, their heads are black, their eyes have white arcs around them, and their bills are red. In the winter, their heads are gray and their bills are black.

Size: 15 to 18 inches (38 to 46 cm) long

Range: North America, Central America, and South America

Habitat: Beaches, salt marshes, crop fields, and landfills

Diet: Earthworms, snails, crabs, insects, fish, berries, and garbage

FUN FACT
Adult laughing gulls remove broken shells from their nests after an egg hatches. They do this so the pieces don't get stuck on other eggs and keep them from hatching.

Winter coloring

Summer coloring

HAMERKOP *(SCOPUS UMBRETTA)*

The hamerkop is also known as the hammerhead stork. That's because the clump of feathers on the back of its head, together with its bill, looks like a hammer. This bird is usually found alone or in pairs while wading in water for food. Its bill, legs, and feet are black. Hamerkops use their feet and wings to flush out prey, which they then grab with their bills. They also use their bills to probe the mud for food.

FUN FACT
The hamerkop often rides on the backs of hippos to search the ground for frogs.

HOW TO SPOT

Color Pattern: Hamerkops are brown all over with a purplish gloss on their backs. They have darker bands on their upper tails and a crest on the back of their heads.

Size: 19 to 22 inches (48 to 56 cm) long

Range: Sub-Saharan Africa and Madagascar

Habitat: Wetlands, including freshwater marshes; edges of lakes and rivers

Diet: Frogs, small fish and mammals, crustaceans, and worms

SADDLE-BILLED STORK
(EPHIPPIORHYNCHUS SENEGALENSIS)

The saddle-billed stork is thought to be the world's tallest stork. Males are larger than females, but they have the same coloring. They can be told apart by their eyes: females have yellow eyes and males have dark brown. The saddle-billed stork's beak is red and black with a yellow patch on top that looks like a saddle.

HOW TO SPOT

Color Pattern: Their bodies are white with a bare, red patch in the center of their breasts. Their heads, necks, and tail feathers are black, and their flight feathers are white.

Size: 55 to 59 inches (140 to 150 cm) tall; 11 to 16 pounds (5 to 7.2 kg)

Range: Sub-Saharan Africa

Habitat: Open spaces along wetlands and swamps

Diet: Fish, crustaceans, frogs, reptiles, and small mammals

YELLOW-BILLED SPOONBILL
(PLATALEA FLAVIPES)

Yellow-billed spoonbills have a spoon-shaped bill. They wade in water with their yellow feet and legs, sweeping their bills from side to side. When a spoonbill detects an insect, it quickly grabs it. The bird then lifts its bill and lets the food slide down its throat.

HOW TO SPOT

Color Pattern: These birds are white with a yellow face.
Size: 30 to 36 inches (76 to 92 cm) long
Range: Australia
Habitat: Freshwater wetlands, lagoons, swamps, and sometimes dry pastures
Diet: Aquatic insects and their larvae

FUN FACT
The yellow-billed spoonbill has vibration detectors inside its bill. The detectors help the bird find food even in cloudy water or at night.

SPOONBILLS

There are six species of spoonbills in the world, and they all have spoon-shaped bills. The roseate spoonbill is the only spoonbill in the Americas. Most spoonbill species have white feathers, but the roseate's feathers are pink because of the shrimp it eats.

DEMOISELLE CRANE
(ANTHROPOIDES VIRGO)

Demoiselle cranes have long necks and black feathers on their heads. Long, black feathers hang from their lower necks down to their breasts. These cranes feed in crop fields in early morning and early evening. During the hot time of the day, they rest on the edges of marshes. When courting, they perform graceful dances. They bow, run and jump, and throw plants.

HOW TO SPOT

Color Pattern: This bird is gray with a black head and neck. Its ear tufts are white.

Size: 37 inches (95 cm) long

Range: Breeds in central Eurasia and spends winter in India and sub-Saharan Africa

Habitat: Crop fields, sandy riverbanks, and ponds

Diet: Grass seeds, grains, insects, lizards, and worms

BLACK-CROWNED NIGHT HERON *(NYCTICORAX NYCTICORAX)*

Black-crowned night herons feed mostly at night. During the day, they hunch in vegetation at the water's edge. These herons have red eyes. They are the most widespread heron in the world and can live in saltwater and freshwater marshes.

HOW TO SPOT

Color Pattern: They have black backs and caps and whitish-gray underparts. Breeding birds have two long plumes, which are long feathers displayed on the back of the head.

Size: 25 inches (64 cm) long

Range: North America, the Caribbean, and Central America

Habitat: Wetlands, including marshes, swamps, streams, rivers, lakes, and ponds

Diet: A variety, including worms, crayfish, lizards, plant material, birds, and eggs

FUN FACT

Black-crowned night herons nest in trees in groups, with often a dozen nests per tree. These colonies may exist for 50 years or more.

GREAT BLUE HERON
(ARDEA HERODIAS)

Great blue herons stand with their long necks straight or with their heads back on their shoulders. When they fly, their necks are folded in an *S* shape, and their legs trail straight behind. Their long bills are orange, and they usually hunt for food alone. They slowly wade or stand perfectly still in water to search for food. They nest in colonies of several hundred pairs. Their large nests are made from sticks and can be built on the ground, on bushes, or in trees.

HOW TO SPOT

Color Pattern: The bird is blue gray with a black stripe over each eye. Its black crown holds black head plumes.

Size: 38 to 54 inches (97 to 137 cm) long; around 4.5 to 5.5 pounds (2 to 2.5 kg)

Range: North America, South America, Europe, and Africa

Habitat: Saltwater and freshwater marshes, swamps, shores, tidal flats, and wet fields

Diet: Fish, amphibians, reptiles, small mammals, insects, and birds

KILLDEER *(CHARADRIUS VOCIFERUS)*

Killdeer run across the grass in sprints, stopping quickly to check for insect prey. They're named for the *kill-deer* call they repeat. Killdeer often make their calls while circling overhead. They fly stiffly on long, pointed wings. Adult and young killdeer are also good swimmers.

FUN FACT

When a predator gets close to a killdeer's nest, the adult bird pretends to have a broken wing to lead the predator away. Once the predator has been lured away from the nest, the killdeer flies off.

HOW TO SPOT

Color Pattern: The killdeer has brownish-tan upperparts and white underparts. Two black bands cross the bird's white chest. The bird has a brown face with black-and-white patches, and an orange rump is seen in flight.

Size: 7.9 to 11 inches (20 to 28 cm) long

Range: North America

Habitat: Sandbars, mudflats, fields, lawns, parking lots, golf courses, and ponds

Diet: Earthworms, snails, crayfish, beetles, and grasshoppers

THREE-BANDED PLOVER
(CHARADRIUS TRICOLLARIS)

The three-banded plover picks up food about ten to 40 times per minute. The plover uses its feet to find prey in the sand and muck. Three-banded plovers have a pinkish-red bill with a black tip, and brown eyes that are circled in orange. These birds may scare off intruders by making a high-pitched whistle while bobbing their heads.

HOW TO SPOT

Color Pattern: This bird has a dark brown head and back, a brown crown circled by a white stripe, and white underparts. Two black bands separate a third white band on the bird's chest.

Size: 7 inches (18 cm) long

Range: Sub-Saharan Africa

Habitat: Edges of wetlands; muddy ground next to lakes, ponds, and rivers

Diet: Insects and their larvae, small crustaceans, mollusks, and worms

WHAT DO SHOREBIRDS HAVE IN COMMON?

Shorebirds like to be near water, whether it's on the coast, in a marsh, or next to a river. Shorebirds share physical traits such as round heads, long legs, and bills that help them find food in mud and sand. They also usually gather in large flocks that include different species.

AMERICAN AVOCET
(RECURVIROSTRA AMERICANA)

American avocets have a long, black, upturned bill that helps them find and catch food. They find food by sweeping their slightly open bills side to side in the water. They catch food by pecking at it with their pointed bills. Sometimes, they duck their heads underwater to catch prey.

HOW TO SPOT

Color Pattern: Its head and neck are rusty orange while breeding and gray when nonbreeding. It has a white body with a black patch on its back. Its wings are black and white.

Size: 16.9 to 18.5 inches (43 to 47 cm) long

Range: Western United States and Mexico

Habitat: Freshwater and saltwater wetlands, mudflats, and flooded pastures

Diet: Beetles, flies, gnats, water fleas, and small fish

Breeding colors

Nonbreeding colors

FUN FACT

American avocet chicks leave the nest a day after hatching. At that point, they can already walk, swim, and dive underwater.

36

PERUVIAN THICK-KNEE
(BURHINUS SUPERCILIARIS)

Peruvian thick-knees are mostly active at night. During the day they remain still, roosting in open areas that blend with their colors. Because of this, it can be hard to see these birds. It's possible to find them by searching for the thick, three-toed tracks they leave in the sand. They like open crop fields and avoid thick vegetation. Peruvian thick-knees have long, yellowish legs and a yellowish bill with a black tip.

HOW TO SPOT

Color Pattern: These birds are yellowish gray with spots, brown lines, and a white belly. They have a white line over large, round eyes. A black line circles the crown.

Size: 15 to 17 inches (38 to 43 cm) long

Range: South America (Chile, Ecuador, and Peru)

Habitat: Crop fields, dry shrublands, flooded grasslands, and pastures

Diet: Small insects and grains

Dull-colored feathers help the Peruvian thick-knee blend in with its surroundings.

BAR-TAILED GODWIT
(LIMOSA LAPPONICA)

Bar-tailed godwits wade in deep water to search for food and jam their sharp bills deep in the mud. These long bills have black tips and are turned up slightly. Bar-tailed godwits have been known to fly as far as 6,800 miles (11,000 km) without stopping. Like many birds, the male bar-tailed godwit's breeding colors are more vibrant than its nonbreeding colors.

Males and females have similar nonbreeding colors.

Male breeding colors

HOW TO SPOT

Color Pattern: These birds are spotted brown on top with lighter brown underparts. Their underwings are white, and their tails are white with brown bands. Breeding males have a reddish-brown head and chest.

Size: 15 to 18 inches (38 to 46 cm) long

Range: Breed in Scandinavia, northern Asia, and Alaska; migrate to Indonesia, New Guinea, Australia, and New Zealand

Habitat: Mudflats, beaches, and trees in swampy areas

Diet: Mollusks, worms, and water insects

SPOTTED SANDPIPER
(ACTITIS MACULARIUS)

The spotted sandpiper is the most widespread sandpiper in North America. It looks as if it's always leaning forward. This sandpiper bobs its tail up and down as it walks. When looking for food, the spotted sandpiper crouches low. When this bird spots prey, it runs toward it, tail bobbing the whole time. Spotted sandpipers let out high whistles as they take off in flight.

HOW TO SPOT

Color Pattern: Breeding adults have a dark brown back, a white chest with dark spots, and an orange bill.

Size: 7.1 to 7.9 inches (18 to 20 cm) long

Range: North America, Central America, and South America

Habitat: Pebbly shores beside streams, ponds, and marshes

Diet: Flies, grasshoppers, beetles, snails, worms, and small fish

FUN FACT
Unlike other species of migrating birds, the female spotted sandpiper establishes and defends her territory. In other migratory bird species, it's usually the male who does this.

GAMBEL'S QUAIL
(CALLIPEPLA GAMBELII)

Gambel's quail can be seen in a group known as a covey. As many as 12 birds flock together. They scratch the ground under cacti and shrubs for food. When threatened, they run instead of fly. Before the eggs hatch, the female quail calls to her chicks. The chicks make sounds to each other from inside their eggs. All these chicks hatch at the same time.

HOW TO SPOT

Color Pattern: Their backs and chests are gray, and they have red-brown patches on their sides that are striped with white. Both males and females have a black crest. The male's face is black and his belly is cream colored with one black patch. Females are plainer.

Size: 9.8 inches (25 cm) long

Range: United States and Mexico

Habitat: Thorny vegetation in deserts and river valleys

Diet: Seeds, leaves, and grass-blades

Female   Male

40

VULTURINE GUINEA FOWL
(ACRYLLIUM VULTURINUM)

Vulturine guinea fowl have blue-gray skin, and their necks and heads are mostly bald. These birds have short, bushy, brown feathers on the back of their heads. Vulturine guinea fowl have red eyes and a dark gray beak. These birds don't fly often, and when threatened they usually run. Their nests are shallow scrapes in the ground, where females lay up to 12 eggs.

HOW TO SPOT

Color Pattern: They have bright-blue feathers streaked with black and white. They have small, white dots on their tail feathers.

Size: 20 to 24 inches (51 to 61 cm) long

Range: Eastern tropical Africa

Habitat: Deserts with tall grasses or thorn bushes

Diet: Seeds, fruit, roots, grubs, insects, and small reptiles

FUN FACT

Vulturine guinea fowl keep an eye on monkeys high above in the trees. These birds will snatch up any fruit that the monkeys drop.

SOUTHERN SCREAMER
(CHAUNA TORQUATA)

Southern screamers have small heads when compared with their large bodies. During breeding season, these birds are often seen in pairs. The male sits where he can watch for danger as the female looks for food in shallow water. It's common for them to take flight late in the morning. They soar overhead, calling loudly.

HOW TO SPOT

Color Pattern: These birds are gray with a black neckband, white cheeks, and gray chest. The bare skin around their eyes is reddish.

Size: 33 to 37 inches (84 to 94 cm) long; around 9 to 10 pounds (4 to 4.5 kg)

Range: Southern South America

Habitat: Wetlands and open marshes next to grasslands

Diet: Insects, seeds, and other plant material

WILD TURKEY *(MELEAGRIS GALLOPAVO)*

Wild turkeys have long legs, bare skin on their heads and necks, and wide tails. Males have a red wattle. These birds mostly walk, but they can also run and fly. At night they roost in trees. Wild turkeys have many predators, including bobcats, coyotes, birds of prey, and humans. Male wild turkeys do not help raise their young. A female and her chicks form a family group. Multiple groups often flock together.

HOW TO SPOT

Color Pattern: Wild turkeys have dark bodies with a brownish-green tint. Their wings are dark and barred with white. Turkeys in the Rocky Mountains have tail feathers tipped with white.

Size: 43 to 45 inches (109 to 114 cm) long; 5.5 to 24 pounds (2.5 to 11 kg)

Range: North America

Habitat: Forests of oak, beech, and hickory trees

Diet: Plants, seeds, berries, tree nuts, and fruit

FUN FACT

Male wild turkeys make a *gobble* sound that can be heard up to 1 mile (1.6 km) away. They usually make their calls from up in trees, which helps the sound carry.

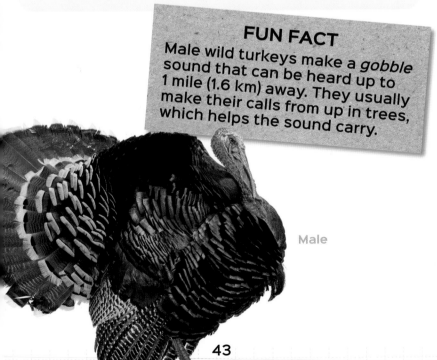

Male

43

RED-TAILED HAWK
(BUTEO JAMAICENSIS)

Red-tailed hawks are the most common hawk species in North America. They soar above open fields, turning in slow circles. From below, their red tails are visible. The underside of their wings is light except for the dark wing tips. They often perch on telephone poles and wires, scanning the ground for prey. Pairs who have mated usually stay together until one dies.

FUN FACT

Red-tailed hawks court by making wide circles as they fly high in the sky. Then, they sometimes grab each other, clasping talons, and spiral toward the ground before letting go.

HOW TO SPOT

Color Pattern: Their upperparts are dark brown and underparts are pale. The tail has a dark rust coloring above and is lighter below.

Size: Around 17.7 to 22.1 inches (45 to 56 cm) long

Range: North America

Habitat: Open areas including deserts, grasslands, fields, and roadsides

Diet: Small mammals, birds, and snakes

TURKEY VULTURE *(CATHARTES AURA)*

Turkey vultures have red heads that lack feathers, and they have pale bills. These birds fly with their wings raised in a *V*, wobbling from side to side. They usually fly at low altitudes while searching for carrion. When threatened, they hiss. They also defend themselves by vomiting stomach acids.

HOW TO SPOT

Color Pattern: These birds have brown upperparts with darker brown on the neck and below.

Size: About 25.2 to 31.9 inches (64 to 81 cm) long; 4.4 pounds (2 kg)

Range: North America, Central America, and South America

Habitat: Open areas including farmland, forests, landfills, and roadsides

Diet: Dead mammals, reptiles, birds, and fish

TURKEY VULTURES DO ALL THE WORK

Unlike other birds of prey, turkey vultures have a powerful sense of smell. Other birds watch as turkey vultures find the food, then chase the vultures off when they see them circling a carcass. Eagles and coyotes get to eat first. Then condors eat. Last in line are ravens and the turkey vultures.

BALD EAGLE
(HALIAEETUS LEUCOCEPHALUS)

When they've reached maturity, bald eagles have a white head and tail. Their bodies are dark brown, and they have yellow eyes and a yellow bill. These large birds have a wingspan of around 6 to 8 feet (1.8 to 2.4 m) long. They also have sharp claws. Bald eagles usually fly alone, but many eagles will sometimes gather in an area that has a lot of prey.

HOW TO SPOT

Color Pattern: Bald eagles have white feathers on their heads and tails, and they have dark brown bodies.

Size: 2.8 to 3.5 feet (0.9 to 1.1 m) long; 6.5 to 14 pounds (2.9 to 6.4 kg)

Range: North America

Habitat: Near large bodies of water

Diet: Fish, waterbirds, small mammals, and carrion

GYRFALCON *(FALCO RUSTICOLUS)*

Fierce gyrfalcons are the largest falcon species in the world. They have yellow rings around their dark eyes, and their hooked bills are gray with yellow at the base. Males are smaller than females. Gyrfalcons hunt large birds, chasing them down in flight. When hunting mammals such as hares, the falcons fly low to the ground. Gyrfalcons may chase a hare for several miles. They can reach speeds of 130 miles per hour (210 kmh).

HOW TO SPOT

Color Pattern: Gyrfalcons are shades of white, gray, and dark brown. Their underparts are whitish spotted with gray.

Size: 18.9 to 25.2 inches (48 to 64 cm) long; 1.8 to 4.6 pounds (0.8 to 2 kg)

Range: Northern United States, Alaska, and Canada

Habitat: Rocky seacoasts, offshore islands, and river bluffs

Diet: Birds and mammals

FUN FACT

When the chicks can't finish a meal, female gyrfalcons store the leftovers away from the nest. Later, they either eat the food themselves or bring it back for the chicks to finish off.

47

BURROWING OWL
(ATHENE CUNICULARIA)

Burrowing owls have long legs and hunt on the ground in the daytime. Their homes are burrows. They can dig their own but will also live in abandoned prairie dog burrows. These owls nest in colonies with as many as 12 pairs living close together. When they're not hunting, burrowing owls often stand motionless at the entrance to their burrows.

HOW TO SPOT

Color Pattern: Their upperparts are brown and spotted. Their breasts are also spotted, and their bellies have dark bars. Their throats and eyebrows are white.

Size: 7.5 to 9.8 inches (19 to 25 cm) long

Range: Western Canada and United States, Mexico, and Central America

Habitat: Deserts, grasslands, pastures, crop fields, and golf courses

Diet: Insects, birds, lizards, snakes, and small mammals

GETTING A GOOD VIEW

An owl's eyes stay in one place—they don't move! That means if an owl wants to see from side to side, it must move its head. Owls can twist their necks quite far in each direction. Some owls can turn their heads around 270 degrees. In comparison, humans can only turn their heads about 90 degrees.

GREAT HORNED OWL
(BUBO VIRGINIANUS)

The large great horned owl is one of North America's most common owls. As a strong predator, it can kill larger birds and mammals. These owls use their powerful talons to snap the spines of their prey. They hunt mostly at night and sometimes at dusk. They scan for prey from a perch and then swoop down for the kill.

HOW TO SPOT

Color Pattern: This owl is a spotted gray brown. Its face is reddish brown, the throat has a white patch, and its ears have large feather tufts.

Size: 18.1 to 24.8 inches (46 to 63 cm) long; 2 to 5 pounds (1 to 2.3 kg)

Range: North America and Central America

Habitat: Forests, fields, wetlands, and pastures

Diet: Mammals, birds, insects, reptiles, and fish

FUN FACT
Great horned owls have the biggest eyes of any owl.

MOURNING DOVE
(ZENAIDA MACROURA)

Mourning doves have black eyes, red feet, and long, pointed tails. Their wings make a trembling whistle sound when they take flight. Seeds make up most of their diets. They feed by pecking on the ground, swallowing many seeds to fill a pouch in their throats before flying off to perch and digest the food.

HOW TO SPOT

Color Pattern: This dove is tan overall with black spots on its wings. Its tail feathers are long, dark, and pointed.

Size: 9.1 to 13.4 inches (23 to 34 cm) long

Range: North America

Habitat: Grasslands, crop fields, backyards, and roadsides

Diet: Seeds

HOW DO BIRDS AVOID COLLISIONS?

Scientists in Australia were curious as to why birds don't crash into each other while flying. They did an experiment with an air tunnel and parakeets. As two birds flew toward each other, both parakeets instinctively veered to their right. Sometimes they also changed altitude so that one flew higher and the other lower. The scientists videotaped birds flying toward each other 102 times. The parakeets made all those flights without one collision.

COMMON PIGEON *(COLUMBA LIVIA)*

Common pigeons are also called rock doves. They can have red-orange eyes and red feet. Pigeons are a common sight in cities because they rely on humans for food and shelter. City pigeons build nests on buildings and window ledges, while pigeons in the country nest in barns and under bridges. Pigeons eat by pecking at food on the ground. They use their bills like straws to drink water.

HOW TO SPOT

Color Pattern: Colors can vary, but the most common is blue gray with two black bands on the wings. They have a black-tipped tail, and the feathers on their throats are multicolored.

Size: 11.8 to 14.2 inches (30 to 36 cm) long

Range: North America and cities around the world

Habitat: Towns, cities, farmland, and rocky cliffs

Diet: Seeds, fruit, and crumbs left by humans

FUN FACT
Pigeons can find their way home even when blindfolded. They can sense Earth's magnetic field, which helps them navigate.

51

TAWNY FROGMOUTH
(PODARGUS STRIGOIDES)

Tawny frogmouths are nocturnal birds with big, yellow eyes. They are often mistaken for owls, but they are part of the nightjar family. During the day their colorings help them blend in with trees. At night they perch on branches, waiting for prey. Then they pounce and use their wide, hooked beaks to kill. Frogmouths aren't very talkative. When threatened, they hiss.

HOW TO SPOT

Color Pattern: Overall, these birds are silver gray with black streaks and spots. Their underparts are slightly paler.

Size: 12.9 to 19.6 inches (33 to 50 cm) long

Range: Australia and Tasmania

Habitat: Forests, edges of rain forests, parks, and gardens with trees

Diet: Insects, slugs, worms, snails, small mammals, reptiles, frogs, and birds

FUN FACT
Frogmouths sleep on tree branches during the day. If discovered, they stiffen their bodies and pretend to be part of the branch.

The tawny frogmouth's coloring helps to camouflage it.

MEXICAN WHIP-POOR-WILL
(ANTROSTOMUS ARIZONAE)

The Mexican whip-poor-will is a shy night bird. It flies with its wide mouth open to catch insects and then swallows them whole. Mexican whip-poor-wills roost on the ground. They also lay eggs on the bare ground rather than in a nest. During the night, males will sing to find a mate and to defend their territories. They get their name from their song, which sounds like *whirr-p-wiirrr*.

HOW TO SPOT

Color Pattern: Their upperparts are gray brown and spotted with black, pale gray, and tan. Their underparts are lighter and barred with fine, black lines. The throat is black, and the collar is white.

Size: 9 to 10.4 inches (23 to 26 cm) long

Range: Southwestern United States and Mexico

Habitat: Mountains, in pine and oak trees

Diet: Moths, beetles, and mosquitoes

HAIRY WOODPECKER
(PICOIDES VILLOSUS)

The hairy woodpecker searches for food on the trunks and limbs of trees, and also on fallen logs. This bird often removes bark with its beak to go after insects. A hairy woodpecker can be located by the sound of its tapping beak. It moves up the tree trunk by propping itself with stiff tail feathers and then leaping with both feet.

HOW TO SPOT

Color Pattern: This bird has black wings with white spots, and its head is black with two white stripes. A male will have a red spot on the back of its head.

Size: 7.1 to 10.2 inches (18 to 26 cm) long

Range: North America and Central America

Habitat: Forests, forest edges, orchards, and near beaver ponds

Diet: Insects, especially ants and bark beetle larvae

Male

Female

54

NORTHERN FLICKER
(COLAPTES AURATUS)

The northern flicker is a large woodpecker. It can be identified by the white patch on its rump that's shown in flight. Unlike other woodpeckers, northern flickers mostly search for food on the ground. They dig for insects with their slightly curved bills and use their long tongues to lick up ants. However, they communicate like other woodpeckers, drumming on objects to defend their territories.

HOW TO SPOT

Color Pattern: This bird is brown with black spots and bars, and it has a white patch on its rump. It also has a black bib and a red patch on the back of its head. Underneath, the wings and tail are bright yellow in eastern birds and red in western birds. Males in the east have a black "whisker," while males in the west have a red one.

Size: 11 to 12.2 inches (28 to 31 cm) long

Range: North America

Habitat: Woodlands, forest edges, and open fields with trees

Diet: Insects

Eastern male

FUN FACT

Northern flickers fly like other woodpeckers. They flap hard and glide between flaps, moving in an up-and-down pattern.

55

CUBAN TROGON
(PRIOTELUS TEMNURUS)

The red-eyed Cuban trogon is the national bird of Cuba. Its red, white, and blue feathers share the colors of Cuba's national flag. Its upper bill is dark, and the lower part is red. These birds hover, continuing to flap their wings as they pause in mid-flight to feed on vegetation.

HOW TO SPOT

Color Pattern: Cuban trogons have a purple-blue crown and neck. The throat and breast are white, and the belly is red.

Size: 9 to 9.8 inches (23 to 25 cm) long

Range: Cuba

Habitat: Damp and dry forests; mountainous areas

Diet: Mostly flowers; also buds, fruit, and insects

ELEGANT TROGON
(TROGON ELEGANS)

The elegant trogon is a brightly colored bird with a long tail and yellow bill. It sits motionless in trees as it scans for fruit or prey. Once it spots something, it bursts into flight. Elegant trogons can't drill their own holes in trees, so they rely on woodpeckers for nests. Once the woodpeckers have moved on, the trogons move in.

FUN FACT
Of all trogons, the elegant trogon can live in the widest variety of habitats. For example, in Guatemala it can be found from sea level to 6,200 feet (1,900 m) above sea level.

Female

Male

HOW TO SPOT

Color Pattern: Males have green coloring on their heads and upperparts. They also have a black face and throat, and a red belly with a white chest band. Females have a gray head and chest, a pale red lower belly, and a white teardrop by their eyes. Both sexes have a long tail with bands of black and white on the underside.

Size: 12.5 inches (32 cm) long

Range: Arizona, Mexico, and Central America

Habitat: Mountain ranges, canyons, and forests

Diet: Insects and fruit

AMERICAN PYGMY KINGFISHER *(CHLOROCERYLE AENEA)*

The tiny American pygmy kingfisher looks like other kingfishers, sharing their short tails and long bills. It likes to perch on low branches close to streams. It plunges into the water headfirst to catch small fish. These birds also fly after insects, twisting and turning in the air. The American pygmy kingfisher nests in burrows along riverbanks.

Male

Female

HOW TO SPOT

Color Pattern: This bird's upperparts are green with a yellow-orange collar. The underparts are red brown with a white belly. Females have a green breastband.

Size: 5.1 inches (13 cm) long

Range: Southern Mexico, Central America, and parts of South America

Habitat: Dense forests next to streams and rivers with vegetation along banks

Diet: Small fish, tadpoles, and insects

LAUGHING KOOKABURRA
(DACELO NOVAEGUINEAE)

The laughing kookaburra is the largest kingfisher in the world. It smashes larger prey such as snakes and lizards on a hard surface. This kills and softens the prey for eating. These kingfishers plunge into water for prey and also to bathe. The laughing kookaburra is named for its call, a *kook-kook-kook* sound that gets louder and then slows to a chuckle. They often call as a group.

HOW TO SPOT

Color Pattern: The bird's back, wings, and crown are brown. The wings also have pale blue spots.

Size: Up to 18 inches (46 cm) long

Range: Australia, including Tasmania

Habitat: Woodlands, forest clearings, farmland, and orchards

Diet: Insects, reptiles, mammals, and crustaceans

FUN FACT
The call of the laughing kookaburra can get so noisy that the sounds can be mistaken for donkeys or monkeys.

NORTHERN CARMINE
BEE-EATER *(MEROPS NUBICUS)*

Northern carmine bee-eaters gather in flocks above herds of grazing animals. That way, they can catch the insects that fly up from underfoot. These birds also like to ride on the backs of elephants, donkeys, or larger birds for the same reason. They nest in holes in riverbanks, usually in colonies. Sometimes more than 10,000 birds form one colony. These birds have black bills and red eyes.

HOW TO SPOT

Color Pattern: The bird has a blue-green head and olive-green throat. The upperparts and belly are bright red. The rump and underneath the tail are blue green.

Size: 9.4 to 10.6 inches (24 to 27 cm) long

Range: Sub-Saharan Africa

Habitat: Bushy and wooded plains, rivers, marshes, and riverside cliffs

Diet: Insects

FUN FACT

The northern carmine bee-eater does eat some bees. But unlike other bee-eaters, it mostly likes grasshoppers and locusts.

EUROPEAN BEE-EATER
(MEROPS APIASTER)

While they do enjoy bees, European bee-eaters eat more than 300 kinds of invertebrates. These birds will swoop into a web to snatch a spider. They hunt on their own but nest in colonies of eight pairs. When they catch stinging insects, they grab them by the middle and fly back to their perches, where they smack the stinger against the branch to rip it out.

HOW TO SPOT

Color Pattern: The bird's crown and neck are dark brown, turning lighter on the back. Its chin and throat are yellow, its chest and belly are blue, and the tail is mostly blue.

Size: 10.6 to 11.8 inches (27 to 30 cm) long

Range: Africa and much of Europe

Habitat: Near fresh waters in forests, grassy plains, and farmland; nests in riverbanks

Diet: Bees, dragonflies, and other flying insects

WHERE DO BIRDS GET THEIR COLORS?

There are two possible sources for the colors of birds' feathers. The first is pigmentation. Pigments are colored materials in plants and animals. One type of pigment in plants makes goldfinches yellow and cardinals red after they eat those plants. The second source for birds' colors is light refraction. Birds such as blue jays have tiny air pockets in their feathers that scatter the light, making the color blue.

LILAC-BREASTED ROLLER
(CORACIAS CAUDATUS)

The rainbow-colored lilac-breasted roller perches on dead trees to watch for prey. These birds often swoop down to catch their prey and eat it on the ground. Sometimes, they return to their perches and smash the prey to death before swallowing it whole. The lilac-breasted roller is known for hunting small animals as they run from bushfires. Their call is a harsh, squawking *zaaak*. These birds have a black bill and brown eyes.

FUN FACT

Rollers get their name from their courtship flights. They dive from high in the air, making a rolling motion as they call out loudly.

HOW TO SPOT

Color Pattern: A lilac-breasted roller has a light green head, a brown back, a violet rump, a whitish chin, and a light purple breast. Its underparts are green blue.

Size: 14.5 inches (37 cm) long

Range: Eastern and southern Africa

Habitat: Grasslands, open woods, and isolated palm trees

Diet: Grasshoppers, beetles, lizards, and small amphibians

INDIAN ROLLER
(CORACIAS BENGHALENSIS)

Indian rollers are often seen perched on bare trees or wires, and dropping to the ground to catch their prey. They go where there are fires and also follow tractors to find any insects revealed by the machines. To bathe, these birds dive in the water from high up. They don't fly in flocks, but they do form family groups. They communicate with a call that sounds like *chack* and also make *boink* sounds.

HOW TO SPOT

Color Pattern: It has pale green-brown upperparts, a pink-brown breast, and its wings and tail show bright blue when spread in flight.

Size: 10.2 to 10.6 inches (26 to 27 cm) long

Range: Tropical southern Asia

Habitat: Open grasslands and lightly forested areas

Diet: Insects, spiders, small reptiles, and amphibians

RED-BILLED STREAMERTAIL
(TROCHILUS POLYTMUS)

The stunning red-billed streamertail is a hummingbird and the official bird of Jamaica. The male's long tail feathers, known as streamers, make a whining sound in flight. The birds hover to feed from flowers or sometimes perch on the blooms. They stick their long, red beaks into flowers and then rapidly lick nectar with their long tongues. They can lick up to 13 times per second.

HOW TO SPOT

Male

Color Pattern: Both males and females have a green body and black head. Adult males have black streamers, which are long tail feathers.

Size: 3.25 to 3.5 inches (8.2 to 9 cm) long without the tail

Range: Jamaica

Habitat: All habitats, from man-made environments to forests

Diet: Nectar and insects

FUN FACT

Male and female red-billed streamertails like to take water baths. Males also take sunbaths. They spread their wings wide and hold them open to the sun for several minutes.

BLUE-THROATED HUMMINGBIRD
(LAMPORNIS CLEMENCIAE)

The blue-throated hummingbird is the largest hummingbird north of Mexico. The birds' size helps them dominate at feeders and other food sources. They beat their wings about half as much as other hummingbirds, but still beat them 23 times per second when hovering. They nest off the ground, and females sometimes build their nests on top of other bird species' nests.

Male

Female

HOW TO SPOT

Color Pattern: A male's upperparts are greenish, underparts are gray, and throat is blue. Females are green brown.

Size: 4.3 to 4.7 inches (11 to 12 cm) long

Range: Southwestern United States and Mexico

Habitat: Forests in canyons

Diet: Nectar, flying insects, and spiders

OLIVE-BACKED SUNBIRD
(CINNYRIS JUGULARIS)

The olive-backed sunbird's thin, curved bill is made for getting nectar. These birds sometimes hover next to trees and shrubs, probing their flowers for nectar. They mostly perch on the vegetation to eat. They also grab insects from leaves or from spiderwebs. After laying two eggs, the female has all the responsibility for the young.

Female

FUN FACT

While sunbirds and hummingbirds share similar behaviors and have long, thin bills, the two are not related. Sunbirds are actually songbirds.

Male

HOW TO SPOT

Color Pattern: Both sexes have a brown-green back and yellow underparts. Males have dark blue-purple throats, and females have yellow throats and eyebrows.

Size: 4.5 inches (11 cm) long

Range: Asia to northeastern Australia

Habitat: Areas with trees

Diet: Nectar and invertebrates

BEAUTIFUL SUNBIRD
(CINNYRIS PULCHELLUS)

Beautiful sunbirds eat nectar from the flowers of 20 different plant groups. Males defend their territories and chase invaders away. Instead of inserting its bill into a flower to get the nectar, a beautiful sunbird will sometimes use its sharp bill to poke a hole in the bottom of the flower.

Male

HOW TO SPOT

Color Pattern: A male's head, throat, and back are shiny green. Its chest band is yellow and red, and it has long tail streamers. Females are not colorful. Their upperparts are a dull green brown and underparts are yellowish.

Size: 3.9 inches (10 cm) long

Range: Sub-Saharan tropical Africa

Habitat: Open areas with trees and grassy plains

Diet: Nectar and insects

POLLINATORS

When hummingbirds and sunbirds insert their long, thin bills into flowers to get nectar, pollen sticks to their bills. When the hummingbird or sunbird visits another flower, those pollen grains are left in the new flower. If the two plants are the same kind, pollination happens.

COMMON SWIFT *(APUS APUS)*

Common swifts have long, pointed wings and a forked tail. These birds spend most of their lives in the air except when nesting. They even sleep and mate while flying. Their tiny feet and legs are weak, so they can't perch or walk well. They nest in holes in old buildings. Swifts are often seen flying low and fast around buildings, calling loudly.

HOW TO SPOT

Color Pattern: Overall, the common swift is dark with a white throat.
Size: 6.5 inches (16.5 cm) long
Range: Europe, Africa, and Asia
Habitat: The air, except when nesting
Diet: Flying insects

AUSTRALIAN SWIFTLET
(AERODRAMUS TERRAEREGINAE)

Australian swiftlets roost and breed in caves. These birds form large flocks in and around the caves, with hundreds of swiftlets in a colony. They use echolocation to find ways inside of the dark caves. They also make a low, clicking sound when approaching their nests, probably to warn away other birds.

HOW TO SPOT

Color Pattern: This bird is gray with dark brown upperparts and light gray underparts. Its rump is pale.

Size: 4.3 to 4.7 inches (11 to 12 cm) long

Range: Queensland (a state in Australia)

Habitat: Areas around the tropical coast and offshore islands

Diet: Insects and floating spiders

FUN FACT

The Australian swiftlet's nest is a basket of dried saliva held together with dry grass, twigs, and feathers. They build their nests on cave walls and ceilings.

LOST BIRDS?

Sometimes a bird shows up outside its usual range. One theory about why this happens is that the bird got confused about the direction it was flying during migration. Another theory is that the bird was blown off track during a storm or hurricane. A third idea is that the bird might go to new areas to check them out and see if it can live there.

BARN SWALLOW *(HIRUNDO RUSTICA)*

Barn swallows have a long, deeply forked tail and can fly just inches above the ground or water. Often, they'll dip down for a sip of water or touch their bellies to the water for a quick bath. They eat while in flight. These birds mainly nest in buildings made by humans.

FUN FACT

Barn swallows often get help from other birds while feeding their young. The helpers are usually the chick's older siblings, but sometimes unrelated birds assist them.

Male

HOW TO SPOT

Color Pattern: A male's back, wings, and tail are deep blue. His underparts are tan, his throat and forehead are orange brown, and his white undertail is visible in flight. Female markings are similar but less colorful.

Size: 5.9 to 7.5 inches (15 to 19 cm) long

Range: North America, Central America, and South America

Habitat: Open areas such as parks and farmland; over lakes and ponds

Diet: Flies, beetles, and other flying insects

PURPLE MARTIN *(PROGNE SUBIS)*

The purple martin is the largest swallow in North America. In the eastern United States, these birds nest in birdhouses provided by humans. In the west, they often nest in tree holes made by other birds. Purple martins have a slightly hooked, black bill. These birds capture insects by flaring their tails.

HOW TO SPOT

Color Pattern: They have blue-purple coloring with brownish-black wings and tail. A female's belly and collar are gray.

Size: Around 7.5 to 7.9 inches (19 to 20 cm) long

Range: North America, Central America, and South America

Habitat: Towns, cities, parks, open fields, and streams

Diet: Flying insects

Male   Female

AFRICAN PARADISE FLYCATCHER *(TERPSIPHONE VIRIDIS)*

The brightly colored African paradise flycatcher is very vocal, and its raspy call is often heard before the bird is seen. It likes to sing as it perches in trees. Despite their name, African paradise flycatchers eat more than flies. They catch moths and butterflies in midair. During breeding season, males have orange-brown streamers that can be up to 12 inches (30 cm) long.

FUN FACT

The male African paradise flycatcher does a courtship dance for several females at once. He bobs as he flies, showing off his bright underfeathers and long streamers.

HOW TO SPOT

Color Pattern: The African paradise flycatcher's head, neck, and underparts are black, while the bird's tail and wings are orange brown.

Size: 6.7 inches (17 cm) long

Range: Sub-Saharan Africa

Habitat: Open forests and grassy plains

Diet: Insects

VERMILION FLYCATCHER
(PYROCEPHALUS RUBINUS)

Male vermilion flycatchers are bright red, while females are duller in color. They perch to wait for insect prey. In fact, the male vermilion flycatcher perches for 90 percent of his day. They often perch low in trees, pumping or wagging their tails. Their call is a high *teek*. When the male wants to mate, he brings a butterfly to the female. The female lays two to four eggs in a loose cup nest of grass and twigs.

HOW TO SPOT

Color Pattern: Males have a red head, throat, and belly. They also have an eye mask and a brown neck, back, and wings. Females are gray brown above and pale underneath with a white chest and throat.

Size: 4.8 to 5.4 inches (12.2 to 13.7 cm) long

Range: Southwestern United States, Mexico, and Central America

Habitat: Open woodlands, parks, and areas near water

Diet: Insects, spiders, and crustaceans

Female

Male

EURASIAN JACKDAW
(CORVUS MONEDULA)

The Eurasian jackdaw, also known as the western jackdaw, is a very smart crow. It knows to look for food in parks and garbage dumps. It also catches ants in the air when those insects swarm. In farmland areas, these jackdaws pull ticks off the backs of sheep. They also take wool from the sheep to line their nests. Eurasian jackdaws have grayish-white eyes, and their bills, legs, and feet are black.

FUN FACT
Eurasian jackdaws steal food from other birds. Jackdaws chase the other birds until they drop their prey.

HOW TO SPOT

Color Pattern: The Eurasian jackdaw's upper and underparts are dark gray. Its crown is blackish, and its head and neck are light gray.

Size: 13.3 to 15.3 inches (34 to 39 cm) long

Range: Europe, eastern Asia, and northwestern Africa

Habitat: Open country, farmland, parks, and towns

Diet: Grains, seeds, berries, and invertebrates

PIED CROW *(CORVUS ALBUS)*

Pied crows eat a variety of food, including dead animals. Their strong bills are good for cracking shells and digging up grains. Pied crows catch birds and bats in flight, and they steal eggs from nests. They tap the eggs against a branch to crack them open.

HOW TO SPOT

Color Pattern: The bird is an overall shiny black with a white collar across its breast and belly.

Size: 20 inches (50 cm) long

Range: Sub-Saharan Africa and western Asia

Habitat: Open country near towns and villages

Diet: Seeds, insects, frogs, small rodents, birds, bats, and roadkill

EURASIAN JAY
(GARRULUS GLANDARIUS)

The Eurasian jay's coloring varies according to region, but their harsh call of *aaaack-aaaack* helps identify them. They also imitate other birds' calls and can sound like hawks and owls. The Eurasian jay is shy and scares easily, flapping its wings in a slow, jerky manner to get away. Pairs mate for life but separate during the winter, rejoining in the spring.

HOW TO SPOT

Color Pattern: Overall, the bird is pinkish brown with a white throat and rump. It has white patches on its face and a black mustache stripe. Its wings have a white stripe with black and blue markings. Its tail is black.

Size: 14 inches (35 cm) long

Range: Europe, northwestern Africa, and Asia

Habitat: Forests, parks, gardens, and yards

Diet: Nuts, fruit, insects, eggs, frogs, and small mammals

GREEN JAY *(CYANOCORAX YNCAS)*

The green jay gets food by moving through trees, searching the leaves for prey. It also drops to the ground for food and cracks open nuts by pounding them with its strong bill. Green jays can be seen near wildfires, preying on small animals running from the fire. While there, they hold open their wings so smoke gets in their feathers and kills off lice.

HOW TO SPOT

Color Pattern: A green jay's crown is blue, and its face is blue and black. The throat and breast are black. The back is green, the belly is yellow green, and the bird's outer tail feathers are yellow.

Size: 10.5 inches (27 cm) long

Range: Southern Texas, Mexico, and Central America

Habitat: Brush and woodlands

Diet: Seeds, nuts, insects, spiders, small rodents, eggs, and small birds

FUN FACT

Depending on the region, some young green jays stay with their parents and help raise new chicks before leaving to find their own territories.

BLACK-BILLED MAGPIE
(PICA HUDSONIA)

When hopping on the ground, the black-billed magpie holds its long tail feathers up. In flight, those long tail feathers plus bright-white wing patches make the black-billed magpie easy to identify. They gather in flocks and work together to attack predators. Males perch at the tops of trees to establish their territories and, during courtship, spread their tails. Pairs mate for life.

FUN FACT

If a black-billed magpie finds a magpie that has died, it calls loudly to others. As many as 40 gather to call out for about 15 minutes before they silently fly away.

HOW TO SPOT

Color Pattern: The head, upper breast, back, rump, and tail are black with a green gloss. These birds have a white patch around their bellies. Their folded wings have a white stripe and dark blue areas.

Size: 19 inches (48 cm) long

Range: Western United States

Habitat: Meadows, grasslands, and plains

Diet: Fruit, grains, grasshoppers, beetles, small mammals, eggs, birds, and carrion

RED-BILLED BLUE MAGPIE
(UROCISSA ERYTHRORYNCHA)

The red-billed blue magpie searches for food in trees and on the ground. The bird's legs and feet are red orange, and its eyes are brown. The male courts the female by showing off his wings and 17-inch- (43 cm) long tail with short, bouncy flights. They mate for life and defend their territories, nesting in trees and bushes. Both parents care for their young. These magpies do not migrate except to move to lower elevations in the winter.

HOW TO SPOT

Color Pattern: The bird's upperparts are blue, and its head, neck, and breast are black. Its tail is blue with white tips.
Size: 19.6 to 27.5 inches (50 to 70 cm) long
Range: South Asia and China
Habitat: Forests and farmland
Diet: Insects, invertebrates, fruit, seeds, eggs, and young birds

BIRDS PLANTING TREES

Crows and jays hide seeds and nuts to eat later. These birds are smart, but they don't always remember their hiding places. When those seeds aren't found and eaten, they sprout. In this way, the birds help trees spread their seeds and help forests grow.

RIFLEMAN *(ACANTHISITTA CHLORIS)*

The rifleman is New Zealand's smallest bird. It searches for food by flying in short bursts from tree to tree. These birds mostly hunt prey in the treetops and on tree trunks, and sometimes on the ground. Mates search for food together, the whole time calling out to stay in contact. Parents join with others to form groups and raise their young together.

Male

HOW TO SPOT

Color Pattern: Males have bright-green upperparts and a yellow-green rump. Their underparts are white, and their wings have green, white, and yellow patches. The tail is black with a white tip. Females are a dull brown.

Size: 2.7 to 3.5 inches (7 to 9 cm) long

Range: New Zealand

Habitat: Forests

Diet: Insects, spiders, and fruit

FUN FACT

When feeding their young, the male rifleman searches the leaves while the female hunts on the tree trunk. His green feathers and her brown coloring keep them both hidden.

Female

BLUE-GRAY GNATCATCHER
(POLIOPTILA CAERULEA)

The blue-gray gnatcatcher is a tiny bird with a long tail. It busily searches for food. The bird stays in motion, hopping around vegetation and swishing its tail to scare up insects. Despite its name, these birds don't eat many gnats. Pairs work together to defend their territory. The male sings a soft song, while the female gives an aggressive call.

HOW TO SPOT

Color Pattern: A male's upperparts are blue gray and underparts are gray white. His tail is black with white below and on the edges, and his eyes are black with a white eye-ring. A female's markings are the same, but her upperparts are gray.

Size: 3.9 to 4.3 inches (10 to 11 cm) long

Range: United States, Mexico, and northern Central America

Habitat: Wooded areas near water

Diet: Insects, spiders, and other invertebrates

Male

COMMON NIGHTINGALE
(LUSCINIA MEGARHYNCHOS)

Many people believe common nightingales are the best singers of all birds. Their songs are a combination of notes and whistles. These birds often begin singing at dusk. After dark, they hide in thick vegetation and can be difficult to see. Unpaired males sing at night to find a mate and at dawn to defend their territories. They sing louder in cities to overcome all the noise, which can hurt their vocal cords.

HOW TO SPOT

Color Pattern: The bird's upperparts are brown and underparts are pale.
Size: 5.5 to 6.5 inches (14 to 16.5 cm) long
Range: Europe, Asia, and Africa
Habitat: Forests and woodlands
Diet: Insects, worms, and fruit

SINGING AND BIRD BRAINS

According to researchers, birds with bigger parts of the brain that deal with learning are able to sing more complex songs. Common blackbirds and Eurasian skylarks can learn more notes than most birds. But none know as many notes as the common nightingale. They use 1,160 syllables in their songs.

WARBLING VIREO *(VIREO GILVUS)*

The warbling vireo can be recognized by its fast, lively song. It spends much of its time high in the treetops, searching for caterpillars. Males sing during much of the breeding season. They sing to attract a mate and to defend their territories. When feeding their young, one parent stays at the nest until the other returns.

HOW TO SPOT

Color Pattern: Their upperparts are gray with brownish or greenish tones, and their underparts are whitish.

Size: 5.5 inches (14 cm) long

Range: North America

Habitat: Woodlands and near rivers

Diet: Caterpillars, spiders, moths, butterflies, fruit, and berries

FUN FACT

Like many birds, warbling vireos can look and sound different depending on where they live. Birds in one area can have a different size, bill shape, coloring, and song than warbling vireos from other regions.

INDIGO BUNTING *(PASSERINA CYANEA)*

Bright-blue indigo bunting males are often heard singing along roadsides, defending their territories. The plain female is rarely seen and does most of the work caring for the young. Males sometimes have more than one mate. Young indigo buntings learn their songs from nearby males that aren't their fathers.

HOW TO SPOT

Color Pattern: The male is all blue with a deeper blue head and a silver-gray bill. The female is brown with lighter brown streaks on her chest, and she may have blue patches on her wings. She also has a silver-gray bill.

Size: 4.7 to 5.1 inches (12 to 13 cm) long

Range: Eastern North America, Central America, and South America

Habitat: Brushy edges of woods and fields; along streams and roads

Diet: Seeds, berries, buds, and insects

Male

Female

84

BLACK-CAPPED CHICKADEE
(POECILE ATRICAPILLUS)

Black-capped chickadees have a big, round head and tiny body. They love bird feeders and will calmly eat seeds as the feeder swings in the wind. They hop among tree branches and twigs looking for food, often hanging upside down. Their call is *chick-a-dee-dee-dee*, and their song is a whistled *fee-bee*. Both parents care for young.

HOW TO SPOT

Color Pattern: The bird has a black bib and cap, white cheeks, gray upperparts, and its wing feathers are gray and edged with white.
Size: 5.3 inches (13.5 cm) long
Range: Canada and the United States
Habitat: Forests, clearings, suburbs, and parks
Diet: Seeds, berries, insects, and spiders

FUN FACT

Black-capped chickadees hide seeds in different spots to eat in the future. They can recall thousands of locations where they've hidden food.

PAINTED REDSTART
(MYIOBORUS PICTUS)

The painted redstart looks for food in the treetops and on the ground. It hops from branch to branch, searching for insects. But when it sees a flying insect, it will also fly out to catch it in midair. Its song sounds like *cheery-cheery-cheery-chew*. During courtship, the male and female sometimes sing together.

HOW TO SPOT

Color Pattern: The painted redstart has a black head and upperparts. Its wings are black with white patches, and it has white outer tail feathers and a red belly.

Size: 5.1 to 5.9 inches (13 to 15 cm) long

Range: Southwestern United States, Mexico, and Central America

Habitat: Oak canyons and mountain forests

Diet: Insects, including caterpillars, flies, and small beetles

FUN FACT
While searching for food, the painted redstart displays the white patches on its wings and tail feathers. This behavior seems to scare insects out into the open.

EURASIAN BLACKCAP
(SYLVIA ATRICAPILLA)

Eurasian blackcaps are medium-sized birds that sing a loud, warbling song. Males sometimes mimic other birds' songs as part of their own. These birds are common visitors to feeders and are often aggressive toward other birds. During migration, they may travel 124 miles (200 km) per night.

Male

HOW TO SPOT

Color Pattern: This bird has gray-brown upperparts and light-gray underparts. The male's head is gray with a black cap. Females have similar markings but have an orange-brown cap.

Size: 5.1 inches (13 cm) long

Range: Europe and Asia

Habitat: Forests, wooded areas, parks, and gardens

Diet: Insects, fruit, and berries

Female

AUSTRALASIAN PIPIT
(ANTHUS NOVAESEELANDIAE)

Australasian pipits are ground birds that wag their tails up and down as they search for food on the ground. Their brown color helps them stay hidden from predators. They walk and run along the ground and also perch on fences, stumps, and rocks. During courtship flights, the male raises its tail and sings. These birds' feet and bill are pink gray.

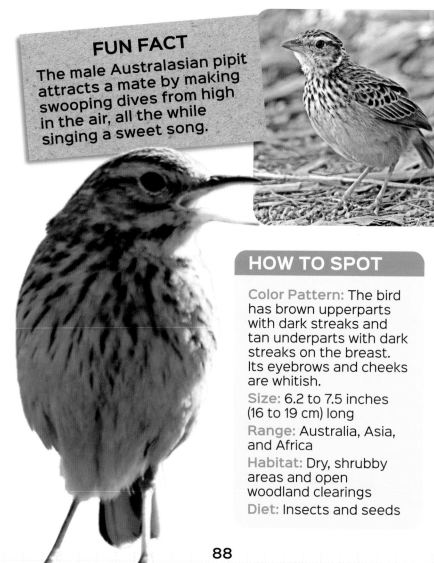

FUN FACT
The male Australasian pipit attracts a mate by making swooping dives from high in the air, all the while singing a sweet song.

HOW TO SPOT

Color Pattern: The bird has brown upperparts with dark streaks and tan underparts with dark streaks on the breast. Its eyebrows and cheeks are whitish.

Size: 6.2 to 7.5 inches (16 to 19 cm) long

Range: Australia, Asia, and Africa

Habitat: Dry, shrubby areas and open woodland clearings

Diet: Insects and seeds

BLUE-AND-WHITE MOCKINGBIRD
(MELANOTIS HYPOLEUCUS)

Blue-and-white mockingbirds often sit in the open at the tops of trees to sing. Their songs are made up of a fast series of chirps, trills, and whistles. They repeat the same song every ten or 20 seconds. Often they imitate other birds' songs—the behavior that gave mockingbirds their name.

HOW TO SPOT

Color Pattern: Its upperparts are blue and underparts are white. This bird has a black mask on its face.
Size: 10 inches (26 cm) long
Range: Mexico and northern Central America
Habitat: Scrubby vegetation and edges of forests
Diet: Insects and small fruit

BIRD SONG VS. CALL

Birds, usually males, sing to claim a territory. During nesting time they sing to defend that territory. Males also sing to get a mate. Female birds do not often sing. But calls are made by both males and females. Calls are shorter and less musical than songs. Birds call out to let others know of a threat and also to locate each other.

NORTHERN CARDINAL
(CARDINALIS CARDINALIS)

Male northern cardinals are bright red. Female northern cardinals only show touches of this color, but they have a crest like the males. These birds are fierce about defending their territories. Both males and females can get so focused on defending against intruders that they will attack their own reflections in a window or mirror.

FUN FACT

The northern cardinal is the state bird of seven US states: Illinois, Indiana, Kentucky, North Carolina, Ohio, Virginia, and West Virginia.

Female

HOW TO SPOT

Color Pattern: Males have a bright-red body, a black face, and a crest. Females are brown with a black face and red bits on wings, tail, and crest.

Size: 8.8 inches (22 cm) long

Range: Eastern United States and Mexico

Habitat: Woodland edges, swamps, and gardens

Diet: Seeds, fruit, and insects

PYRRHULOXIA *(CARDINALIS SINUATUS)*

The pyrrhuloxia, also known as the desert cardinal, lives in the desert and appears to get most of its water from the insects it eats. These birds are territorial during breeding season, but in the winter they flock together to find food. As many as 1,000 birds join a single flock. By early March, the flocks split up, and males sing from perches to establish their territories.

HOW TO SPOT

Color Pattern: Both males and females have a gray body. A male's face, bib, center of breast and belly, and tip of crest are red. Females have a hint of red on their wings and crests.

Size: 8.8 inches (22 cm) long

Range: Southern United States and Mexico

Habitat: Deserts, in thorny brush thickets

Diet: Seeds, fruit, and large insects

Female

Male

CRIMSON-COLLARED GROSBEAK *(RHODOTHRAUPIS CELAENO)*

Both male and female crimson-collared grosbeaks are very colorful. Each has a short, thick, black bill that is curved on top. They have short crown feathers, which they sometimes raise and lower. These birds are often in thick vegetation and can be hard to spot. Their songs are made up of a series of constantly changing notes that rise at the end.

HOW TO SPOT

Color Pattern: Male's underparts and collar are red. Upperparts and wings are blackish. Head, bib, and bill are black. Female markings are similar except underparts and collar are green yellow.

Size: 8.5 inches (21.6 cm) long

Range: Northeastern Mexico and sometimes southern Texas

Habitat: Brushy woodlands and citrus groves

Diet: Leaves and fruit

Female

BEAKS AND SEEDS

A bird's beak is a clue to what that bird eats. Cardinals and grosbeaks have short, thick, cone-shaped beaks fit for eating seeds and nuts. The lower beak fits into a slot in the upper beak. Birds use their tongues to push a seed into the slot and then crush the seed with the lower beak.

BLUE GROSBEAK
(PASSERINA CAERULEA)

Blue grosbeaks often feed on the ground, finding seeds. They also catch insects off vegetation that's low to the ground and can capture insects in flight. During nonbreeding seasons, these birds often form flocks to look for food. The flocks feed in weedy fields. During nesting time, males sing a warbling song to defend their territories.

HOW TO SPOT

Color Pattern: Males have a black face and are overall dull blue. Their wings have two chestnut-colored bars near the shoulders. Females are mostly brown with lighter underparts and some blue on the rump. They don't have black on their faces.

Size: 7.5 inches (19 cm) long

Range: United States, Mexico, and Central America

Habitat: Brush near water

Diet: Insects and seeds

Male

Female

FUN FACT
When a blue grosbeak parent prepares to feed an insect to its young, the parent takes off the wings, head, and many of the legs from the insect.

93

PARADISE TANAGER
(TANGARA CHILENSIS)

Paradise tanagers are brightly colored, lively birds that like to keep moving. They search for insects and fruit in the treetops. They often nest in the outer parts of the tree canopy. They form small flocks of about 15 birds, including birds of other species. Paradise tanagers sing at dawn, repeating *chak-zeet* every two seconds.

HOW TO SPOT

Color Pattern: These birds have a green crown and face, a deep blue throat, and sky-blue underparts. The back of their heads and wings are black, and their rumps are red.

Size: 5 inches (13 cm) long

Range: Northern South America

Habitat: Lowland woodlands and forests

Diet: Berries, fruit, insects, and spiders

WESTERN TANAGER
(PIRANGA LUDOVICIANA)

It's believed that the western tanager gets its red head from a pigment in the insects it eats. These birds stay hidden in the trees but can be heard singing their raspy songs or calling out with a chuckling sound. When they catch dragonflies, they remove the wings and sometimes the heads and legs before eating.

HOW TO SPOT

Color Pattern: Males have a bright-yellow body with a black back. The wings are also black and have a white bar. They have a bright orange-red head. Females have a yellow-green head and body with a grayish back and wings.

Size: 7.3 inches (19 cm) long

Range: Western North America and Central America

Habitat: Forests, wetlands, and parks

Diet: Insects and fruit

Male

FUN FACT

Out of all tanagers, the western tanager is the only one to breed as far north as the Northwest Territories of Canada. Because of the cold, they often stay only two months before flying south again.

Female

95

AMERICAN ROBIN
(TURDUS MIGRATORIUS)

American robins have round bodies and long legs. These birds often run on the ground and then suddenly stop. They tilt their heads and stare at the ground to look for earthworms. Robins sometimes fight over worms. They also perch in trees to sing songs made up of clear, musical whistles.

FUN FACT

Many robins don't migrate, staying in their breeding areas the whole year. During the winter, they spend less time on the ground and more time roosting in trees.

Female

HOW TO SPOT

Color Pattern: Robins have gray-brown upperparts, a dark gray head, and an orange belly. Their throats have black streaks, and they have white on their rumps and undertails. The female's head is lighter gray than the male's.

Size: 7.9 to 11 inches (20 to 28 cm) long

Range: North America

Habitat: Forests, farmland, towns, and lawns

Diet: Insects, berries, and earthworms

GROUNDSCRAPER THRUSH
(PSOPHOCICHLA LITSITSIRUPA)

Groundscraper thrushes use their bills to turn over leaves and find prey. When there's a bushfire, they catch small creatures escaping the fire. They are social and form groups for much of the year. The young from an earlier hatching often help feed new chicks. These birds have a black upper bill and a lower bill that is yellow with a dark tip.

HOW TO SPOT

Color Pattern: Their heads, backs, and wings are grayish. Their faces and underparts are white with black streaks and speckles.

Size: 9.5 inches (24 cm) long

Range: Eastern and southern Africa

Habitat: Mountain grasslands and grassy plains

Diet: Invertebrates

BROWN-HEADED COWBIRD
(MOLOTHRUS ATER)

The brown-headed cowbird got its name because it spends time near grazing cows. The cows scare up insects that the birds can eat. Cowbirds often flock with blackbirds, grackles, and starlings. When the males sing, they lift the feathers on their backs and chests. They also lift their wings, spread their tail feathers, and then take a bow.

HOW TO SPOT

Color Pattern: Both males and females have a brown head. The male's body is black with a greenish shine. Females are brown with streaked underparts.

Size: 7.5 inches (19 cm) long

Range: North America

Habitat: Farms, fields, prairies, and wood edges

Diet: Seeds and insects

Male

DROPPING OFF EGGS

Female brown-headed cowbirds find other birds to take care of their young. The female cowbird lays eggs in another bird's nest. Then, the other bird feeds the young cowbirds, thinking they are their own. Some birds recognize cowbird eggs and remove them. Yellow warblers will build a new nest on top of the cowbird eggs.

98

RED-WINGED BLACKBIRD
(AGELAIUS PHOENICEUS)

Red-winged blackbirds are found throughout North America. Males are often seen perched on a cattail, singing. The females keep out of sight as they search among grasses and reeds for food. A male can have up to 15 females in his territory. They roost in large flocks and then spread out in the morning in search of food. They travel up to 50 miles (80 km) away, then return to flock.

Female

HOW TO SPOT

Color Pattern: Males are black with red patches on their wings and have red feathers bordered by yellow. Females are mostly brown with some white and many dark streaks.

Size: 8.7 inches (22 cm) long

Range: North America

Habitat: Marshes, brushy swamps, hayfields, and mudflats

Diet: Insects and seeds

FUN FACT

Male red-winged blackbirds are very territorial during nesting season. They will chase away nest predators, including horses and humans.

Male

LONG-TAILED MEADOWLARK
(STURNELLA LOYCA)

Long-tailed meadowlarks have a long, thin beak and a flat head. A male meadowlark's song is louder than a female's. His wheezy song starts with a few short whistles. Both sexes use *peet* and *twick* as their calls. Females make cup nests on the ground and often add a roof and tunnel entrance.

HOW TO SPOT

Color Pattern: A male's back and wings are streaked brown. Males have a pink-red belly and chin. Females have similar markings but with more brown coloring, and they only have red on the belly. A female's chin and throat are white.

Size: 9.8 to 10.6 inches (25 to 27 cm) long

Range: Southern South America and Falkland Islands (off the east coast of southern South America)

Habitat: Grasslands, fields, and pastures

Diet: Seeds, insects, and fruit

Female

Male

WESTERN MEADOWLARK
(STURNELLA NEGLECTA)

Western meadowlarks poke at the ground with their bills as they search for insects, grains, and seeds. The western meadowlark's song is a flute-like series of gurgling notes. Males are fierce about protecting their territories and will chase another bird for up to three minutes. The female builds a nest on the ground. If humans get too close to the nest, the birds will abandon it.

HOW TO SPOT

Color Pattern: Their upperparts are brown and black with whitish streaks. They have bright-yellow underparts and a black *V* across their breasts.

Size: 6.3 to 10.2 inches (16 to 26 cm) long

Range: North America

Habitat: Grasslands, fields, meadows, and prairies

Diet: Insects and seeds

FUN FACT

Meadowlarks push their bills into soil or bark and then force it open to make a hole. This helps them reach insects. This technique is called gaping.

HOUSE SPARROW
(PASSER DOMESTICUS)

Male house sparrows have brighter colors than the females. There are many house sparrows in North America. The large population is due to their ability to exist alongside humans. These birds find food in many places—including car bumpers, where house sparrows can feast on smashed insects. They sometimes force other birds from their nests.

HOW TO SPOT

Color Pattern: Males have a gray head and underparts. Their upperparts are streaked with brown, black, and tan. They have a black bib. A female's markings are similar, but her colors are duller and she doesn't have a black bib.

Size: 6.3 inches (16 cm) long

Range: Found nearly all over the world

Habitat: Cities, towns, and farms

Diet: Mostly seeds and some insects

Male

SPARROWS

There are more than 50 sparrow species in North America, and many have brown, spotted feathers. Their colorings help them blend in so predators don't see them. But all that brown can make them difficult to identify. "Little brown job" is what birders often call sparrows they can't identify.

DARK-EYED JUNCO
(JUNCO HYEMALIS)

Dark-eyed juncos search for food by hopping on the ground, pecking and scratching at leaves. They also catch insects by flying low and picking them off vegetation. When dark-eyed juncos fly, they pump their tails and show a bright-white flash of outer feathers.

FUN FACT

When male and female dark-eyed juncos are courting, they jump around with drooping wings. They also spread their tails to display their white outer feathers.

Male

HOW TO SPOT

Color Pattern: Colors vary regionally. Most males have a gray or brown head and breast and white belly, and others are slate colored. Females are browner than males and have lighter markings.

Size: 6.3 inches (16 cm) long

Range: North America

Habitat: Feeders, parks, and open forests

Diet: Seeds, insects, and berries

Female

AMERICAN GOLDFINCH
(SPINUS TRISTIS)

When American goldfinches fly, they often call out *po-ta-to-chip*. They have other calls too, and a male and female pair make the exact same flight call. Other goldfinches can identify members of a pair this way. These small birds balance on thistles and sunflowers as they pluck off the seeds.

FUN FACT
American goldfinches feed only seeds to their young. If a brown-headed cowbird lays eggs in a goldfinch's nest, the young cowbirds won't survive this diet.

Male

HOW TO SPOT

Color Pattern: Males are bright yellow except for a black cap and wings. Females are olive green and pale yellow.

Size: 5 inches (13 cm) long

Range: North America

Habitat: Roadsides, weed and thistle patches, and open woods

Diet: Mostly seeds and some insects

EUROPEAN GOLDFINCH
(CARDUELIS CARDUELIS)

European goldfinches use their narrow bills to dig seeds out of shells. They often hold onto thin plant stems to feed. They will gather in mixed flocks to look for food on the ground. These goldfinches sing a loud, twittering song.

Male

Female

HOW TO SPOT

Color Pattern: Males have a red face, a black crown and collar, and a white throat. Their upperparts are brown, and they have black wings with white spots and yellow bars. Their underparts are whitish. Females have similar markings, but their faces are less red.

Size: 5 inches (13 cm) long

Range: Europe and northern Africa

Habitat: Woodlands and open areas

Diet: Mostly seeds and some invertebrates

ZEBRA FINCH *(TAENIOPYGIA GUTTATA)*

The zebra finch is the most common finch in Australia. These birds feed in large flocks, mainly searching the ground for food. They're noisy, and they have a call that sounds like a toy trumpet. They often perch on fences when not hopping along roadsides in search of seeds. They build large nests that are a bit messy, continuing to add materials after the eggs have been laid.

HOW TO SPOT

Color Pattern: Males have light gray heads and brownish backs. They have orange ear patches, zebra stripes on their chests, white underparts, and black-and-white tails. Females lack the orange patches and stripes on their chests and are instead gray in those places.

Size: 4 inches (10 cm) long

Range: Australia and the Lesser Sunda Islands of Indonesia

Habitat: Scrub forests, plains, and open woodlands

Diet: Grass seeds and some invertebrates

Male

Female

WOODPECKER FINCH
(CAMARHYNCHUS PALLIDUS)

The woodpecker finch doesn't act like most finches. It behaves more like a woodpecker, moving along branches and clinging to tree trunks as it looks for insects. It even pecks at trees like a woodpecker. These birds have a strong bill but don't have a very striking color pattern.

HOW TO SPOT

Color Pattern: Males are overall a sandy-brown color with underparts that are a bit lighter. Their breasts and bellies may be streaked, and their bills are black during the breeding season. Female markings are similar, but they are yellow beige in color and their bills are pale orange.

Size: 5.9 inches (15 cm) long

Range: Galapagos Islands

Habitat: Various, including evergreen and humid forests

Diet: Insects, spiders, and crustaceans

FUN FACT
Woodpecker finches use tools to find food. For instance, some will take a twig and use it to dig out insects.

Female

GLOSSARY

carrion
The rotting flesh of dead animals.

crustacean
A sea animal with an outer skeleton, such as a crab or shrimp.

echolocation
Figuring out the location of something by measuring the time it takes for an echo to bounce back from it.

invertebrate
An animal that does not have a backbone.

light refraction
The separation of white light into colors, as in a prism.

nocturnal
Active at night.

savanna
A flat, grassy plain with few trees.

species
A group of organisms that are very closely related to each other.

sub-Saharan
The part of Africa south of the Sahara Desert.

territorial
Spending time guarding and defending an area that an animal thinks belongs to it.

vegetation
The plants growing in a certain area, such as next to a pond.

vertebrate
An animal with a backbone.

wattle
Loose skin that hangs from the necks and throats of some birds.

TO LEARN MORE

FURTHER READINGS

Beer, Julie. *Birds*. National Geographic, 2016.

Collard, Sneed B. *Fire Birds: Valuing Natural Wildfires and Burned Forests*. Bucking Horse Books, 2015.

Harrison, George H. *Bird Watching for Kids*. Willow Creek Press, 2015.

ONLINE RESOURCES

Booklinks
NONFICTION NETWORK
FREE! ONLINE NONFICTION RESOURCES

To learn more about birds, please visit **abdobooklinks.com** or scan this QR code. These links are routinely monitored and updated to provide the most current information available.

PHOTO CREDITS

ABDOBOOKS.COM

Published by Abdo Publishing, a division of ABDO, PO Box 398166, Minneapolis, Minnesota 55439. Copyright © 2021 by Abdo Consulting Group, Inc. International copyrights reserved in all countries. No part of this book may be reproduced in any form without written permission from the publisher. Abdo Reference™ is a trademark and logo of Abdo Publishing.

Printed in the United States of America, North Mankato, Minnesota.
082020
012021

Editor: Alyssa Krekelberg
Series Designer: Colleen McLaren

Library of Congress Control Number: 2019954298
Publisher's Cataloging-in-Publication Data
Names: Abell, Tracy, author.
Title: Birds / by Tracy Abell
Description: Minneapolis, Minnesota : Abdo Publishing, 2021 | Series: Field guides for kids | Includes online resources and index.
Identifiers: ISBN 9781532193040 (lib. bdg.) | ISBN 9781098210946 (ebook)
Subjects: LCSH: Birds--Juvenile literature. | Birds--Behavior--Juvenile literature. | Animals--Field guides--Juvenile literature. | Reference materials--Juvenile literature.
Classification: DDC 598--dc23